Teacher Certification Exam

Physics Sample Test

Written By:

Leah Sarah Reingold Gordon, MS Physics

To Order Additional Copies:
XAM, Inc.
99 East Central Street
Worcester, MA 01605
Toll Free: 1-800-301-4647
Phone: 1-508-363-0633
Fax: 1-508-363-0634
Email: winwin1111@aol.com
Web: www.xamonline.com

You will find:
- Sample test
- Answers with explanations

XAM, INC.
Building Better Teachers

About the Author

Ms. Gordon teaches physics at Weston High School in Weston, MA, and she has been a high school mathematics and science teacher for eight years. She has enjoyed physics since high school (University High School in Urbana, IL), and pursued that interest in college and graduate school at the Massachusetts Institute of Technology and the California Institute of Technology. When not teaching, Ms. Gordon enjoys playing floor hockey with her husband and two young sons.

Florida: Physics Sample Test
ISBN: 1-58197-091-9

1. A projectile with a mass of 1.0 kg has a muzzle velocity of 1500.0 m/s when it is fired from a cannon with a mass of 500.0 kg. If the cannon slides on a frictionless track, it will recoil with a velocity of _____ m/s.

   A. 2.4

   B. 3.0

   C. 3.5

   D. 1500

2. The weight of an object on the earth's surface is designated x. When it is two earth's radii from the surface of the earth, its weight will be

   A. $x/4$

   B. $x/9$

   C. $4x$

   D. $16x$

3. If a force of magnitude F gives a mass M an acceleration A, then a force $3F$ would give a mass $3M$ an acceleration

   A. A

   B. $12A$

   C. $A/2$

   D. $6A$

4. A car (mass m_1) is driving at velocity v, when it smashes into an unmoving car (mass m_2), locking bumpers. Both cars move together at the same velocity. The common velocity will be given by

   A. m_1v/m_2

   B. m_2v/m_1

   C. $m_1v/(m_1 + m_2)$

   D. $(m_1 + m_2)v/m_1$

5. When acceleration is plotted versus time, the area under the graph represents

A. Time

B. Distance

C. Velocity

D. Acceleration

6. An inclined plane is tilted by gradually increasing the angle of elevation θ, until the block will slide down at a constant velocity. The coefficient of friction, μ_k, is given by

A. cos θ

B. sin θ

C. cosecant θ

D. tangent θ

7. Use the information on heats below to solve this problem. An ice block at 0° Celsius is dropped into 100 g of liquid water at 18° Celsius. When thermal equilibrium is achieved, only liquid water at 0° Celsius is left. What was the mass, in grams, of the original block of ice?

   Given: Heat of fusion of ice = 80 cal/g
                Heat of vaporization of ice = 540 cal/g
                Specific Heat of ice = 0.50 cal/g°C
                Specific Heat of water = 1 cal/g°C

A. 2.0

B. 5.0

C. 10.0

D. 22.5

8. The combination of overtones produced by a musical instrument is known as its

   A. Timbre

   B. Chromaticity

   C. Resonant Frequency

   D. Flatness

9. A long copper bar has a temperature of 60°C at one end and 0°C at the other. The bar reaches thermal equilibrium (barring outside influences) by the process of heat

   A. Fusion

   B. Convection

   C. Conduction

   D. Microwaving

10. The First Law of Thermodynamics takes the form dU = dW when the conditions are

   A. Isobaric

   B. Isochloremic

   C. Isothermal

   D. Adiabatic

11. Given a vase full of water, with holes punched at various heights. The water squirts out of the holes, achieving different distances before hitting the ground. Which of the following accurately describes the situation?

   A. Water from higher holes goes farther, due to Pascal's Principle.

   B. Water from higher holes goes farther, due to Bernoulli's Principle.

   C. Water from lower holes goes farther, due to Pascal's Principle.

   D. Water from lower holes goes farther, due to Bernoulli's Principle.

12. A stationary sound source produces a wave of frequency F. An observer at position A is moving toward the horn, while an observer at position B is moving away from the horn. Which of the following is true?

   A. $F_A < F < F_B$

   B. $F_B < F < F_A$

   C. $F < F_A < F_B$

   D. $F_B < F_A < F$

13. The electric force in Newtons, on two small objects (each charged to –10 microCoulombs and separated by 2 meters) is

   A. 1.0

   B. 9.81

   C. 31.0

   D. 0.225

14. A 10 ohm resistor and a 50 ohm resistor are connected in parallel. If the current in the 10 ohm resistor is 5 amperes, the current (in amperes) running through the 50 ohm resistor is

   A. 1

   B. 50

   C. 25

   D. 60

15. Fahrenheit and Celsius thermometers have the same temperature reading at

   A. 100 degrees

   B. - 40 degrees

   C. Absolute Zero

   D. 40 degrees

16. Which of the following apparatus can be used to measure the wavelength of a sound produced by a tuning fork?

A. A glass cylinder, some water, and iron filings

B. A glass cylinder, a meter stick, and some water

C. A metronome and some ice water

D. A comb and some tissue

17. When the current flowing through a fixed resistance is doubled, the amount of heat generated is

A. Quadrupled

B. Doubled

C. Multiplied by pi

D. Halved

18. The current induced in a coil is defined by which of the following laws?

A. Lenz's Law

B. Burke's Law

C. The Law of Spontaneous Combustion

D. Snell's Law

19. If an object is 15 cm from a convex lens whose focal length is 10 cm, the image is:

A. Virtual and upright

B. Real and inverted

C. Larger than the object

D. Smaller than the object

20. A cooking thermometer in an oven works because the metals it is composed of have different

    A. Melting points

    B. Heat convection

    C. Magnetic fields

    D. Coefficients of expansion

21. In an experiment where a brass cylinder is transferred from boiling water into a beaker of cold water with a thermometer in it, we are measuring

    A. Fluid viscosity

    B. Heat of fission

    C. Specific heat

    D. Nonspecific heat

22. A temperature change of 40 degrees Celsius is equal to a change in Fahrenheit degrees of

    A. 40

    B. 20

    C. 72

    D. 112

23. The number of calories required to raise the temperature of 40 grams of water at 30°C to steam at 100°C is

    A. 7500

    B. 23,000

    C. 24,400

    D. 30,500

24. The boiling point of water on the Kelvin scale is closest to

    A. 112 K

    B. 212 K

    C. 373 K

    D. 473 K

25. The kinetic energy of an object is _____ proportional to its _____.

    A. Inversely...inertia

    B. Inversely...velocity

    C. Directly...mass

    D. Directly...time

26. A hollow conducting sphere of radius R is charged with a total charge Q. What is the magnitude of the electric field at a distance r (given r<R) from the center of the sphere? (k is the electrostatic constant)

    A. 0

    B. $k Q/R^2$

    C. $k Q/(R^2 - r^2)$

    D. $k Q/(R - r)^2$

27. A quantum of light energy is called a

    A. Dalton

    B. Photon

    C. Curie

    D. Heat Packet

28. The following statements about sound waves are true *except*

    A. Sound travels faster in liquids than in gases.

    B. Sound waves travel through a vacuum.

    C. Sound travels faster through solids than liquids.

    D. Ultrasound can be reflected by the human body.

29. The greatest number of 100 watt lamps that can be connected in parallel with a 120 volt system without blowing a 5 amp fuse is

    A. 24

    B. 12

    C. 6

    D. 1

30. A monochromatic ray of light passes from air to a thick slab of glass (n = 1.41) at an angle of 45° from the normal. At what angle does it leave the air/glass interface?

    A. 45°

    B. 30°

    C. 15°

    D. 55°

31. The magnitude of a force is

    A. Directly proportional to mass and inversely to acceleration

    B. Inversely proportional to mass and directly to acceleration

    C. Directly proportional to both mass and acceleration

    D. Inversely proportional to both mass and acceleration

32. A semi-conductor allows current to flow

    A. Never

    B. Always

    C. As long as it stays below a maximum temperature

    D. When a minimum voltage is applied

33. One reason to use salt for melting ice on roads in the winter is that

    A. Salt lowers the freezing point of water.

    B. Salt causes a foaming action, which increases traction.

    C. Salt is more readily available than sugar.

    D. Salt increases the conductivity of water.

34. Automobile mirrors that have a sign, "objects are closer than they appear" say so because

    A. The real image of an obstacle, through a converging lens, appears farther away than the object.

    B. The real or virtual image of an obstacle, through a converging mirror, appears farther away than the object.

    C. The real image of an obstacle, through a diverging lens, appears farther away than the object.

    D. The virtual image of an obstacle, through a diverging mirror, appears farther away than the object.

35. A gas maintained at a constant pressure has a specific heat which is greater than its specific heat when maintained at constant volume, because

    A. The Coefficient of Expansion changes.

    B. The gas enlarges as a whole.

    C. Brownian motion causes random activity.

    D. Work is done to expand the gas.

36. Consider the shear modulus of water, and that of mercury. Which of the following is true?

    A. Mercury's shear modulus indicates that it is the only choice of fluid for a thermometer.

    B. The shear modulus of each of these is zero.

    C. The shear modulus of mercury is higher than that of water.

    D. The shear modulus of water is higher than that of mercury.

37. A skateboarder accelerates down a ramp, with constant acceleration of two meters per second squared, from rest. The distance in meters, covered after four seconds, is

    A. 10

    B. 16

    C. 23

    D. 37

38. Which of the following units is not used to measure torque?

    A. slug ft

    B. lb ft

    C. N m

    D. dyne cm

39. In a nuclear pile, the control rods are composed of

    A. Boron

    B. Einsteinium

    C. Isoptocarpine

    D. Phlogiston

40. All of the following use semi-conductor technology, except a(n):

   A. Transistor

   B. Diode

   C. Capacitor

   D. Operational Amplifier

41. Ten grams of a sample of a radioactive material (half-life = 12 days) were stored for 48 days and re-weighed. The new mass of material was

   A. 1.25 g

   B. 2.5 g

   C. 0.83 g

   D. 0.625 g

42. When a radioactive material emits an alpha particle only, its atomic number will

   A. Decrease

   B. Increase

   C. Remain unchanged

   D. Change randomly

43. The sun's energy is produced primarily by

   A. Fission

   B. Explosion

   C. Combustion

   D. Fusion

44. A crew is on-board a spaceship, traveling at 60% of the speed of light with respect to the earth. The crew measures the length of their ship to be 240 meters. When a ground-based crew measures the apparent length of the ship, it equals

    A. 400 m

    B. 300 m

    C. 240 m

    D. 192 m

45. Which of the following pairs of elements are not found to fuse in the centers of stars?

    A. Oxygen and Helium

    B. Carbon and Hydrogen

    C. Beryllium and Helium

    D. Cobalt and Hydrogen

46. A calorie is the amount of heat energy that will

    A. Raise the temperature of one gram of water from 14.5° C to 15.5° C.

    B. Lower the temperature of one gram of water from 16.5° C to 15.5° C

    C. Raise the temperature of one gram of water from 32° F to 33° F

    D. Cause water to boil at two atmospheres of pressure.

47. Bohr's theory of the atom was the first to quantize

    A. Work

    B. Angular Momentum

    C. Torque

    D. Duality

48. A uniform pole weighing 100 grams, that is one meter in length, is supported by a pivot at 40 centimeters from the left end. In order to maintain static position, a 200 gram mass must be placed _____ centimeters from the left end.

    A. 10

    B. 45

    C. 35

    D. 50

49. A classroom demonstration shows a needle floating in a tray of water. This demonstrates the property of

    A. Specific Heat

    B. Surface Tension

    C. Oil-Water Interference

    D. Archimedes' Principle

50. Two neutral isotopes of a chemical element have the same numbers of

    A. Electrons and Neutrons

    B. Electrons and Protons

    C. Protons and Neutrons

    D. Electrons, Neutrons, and Protons

51. A mass is moving at constant speed in a circular path. Choose the true statement below:

    A. Two forces in equilibrium are acting on the mass.

    B. No forces are acting on the mass.

    C. One centripetal force is acting on the mass.

    D. One force tangent to the circle is acting on the mass.

52. A light bulb is connected in series with a rotating coil within a magnetic field. The brightness of the light may be increased by any of the following except:

    A. Rotating the coil more rapidly.

    B. Using more loops in the coil.

    C. Using a different color wire for the coil.

    D. Using a stronger magnetic field.

53. The use of two circuits next to each other, with a change in current in the primary circuit, demonstrates

    A. Mutual current induction

    B. Dielectric constancy

    C. Harmonic resonance

    D. Resistance variation

54. A brick and hammer fall from a ledge at the same time. They would be expected to

    A. Reach the ground at the same time

    B. Accelerate at different rates due to difference in weight

    C. Accelerate at different rates due to difference in potential energy

    D. Accelerate at different rates due to difference in kinetic energy

55. The potential difference across a five Ohm resistor is five Volts. The power used by the resistor, in Watts, is

    A. 1

    B. 5

    C. 10

    D. 20

56. An object two meters tall is speeding toward a plane mirror at 10 m/s. What happens to the image as it nears the surface of the mirror?

    A. It becomes inverted.

    B. The Doppler Effect must be considered.

    C. It remains two meters tall.

    D. It changes from a real image to a virtual image.

57. The highest energy is associated with

    A. UV radiation

    B. Yellow light

    C. Infrared radiation

    D. Gamma radiation

58. The constant of proportionality between the energy and the frequency of electromagnetic radiation is known as the

    A. Rydberg constant

    B. Energy constant

    C. Planck constant

    D. Einstein constant

59. A simple pendulum with a period of one second has its mass doubled. If the length of the string is quadrupled, the new period will be

    A. 1 second

    B. 2 seconds

    C. 3 seconds

    D. 5 seconds

60. A vibrating string's frequency is _____ proportional to the _____.

    A. Directly; Square root of the tension

    B. Inversely; Length of the string

    C. Inversely; Squared length of the string

    D. Inversely; Force of the plectrum

61. When an electron is "orbiting" the nucleus in an atom, it is said to posses an intrinsic spin (spin angular momentum). How many values can this spin have in any given electron?

    A. 1

    B. 2

    C. 3

    D. 8

62. Electrons are

    A. More massive than neutrons

    B. Positively charged

    C. Neutrally charged

    D. Negatively charged

63. Rainbows are created by

    A. Reflection, dispersion, and recombination

    B. Reflection, resistance, and expansion

    C. Reflection, compression, and specific heat

    D. Reflection, refraction, and dispersion

64. In order to switch between two different reference frames in special relativity, we use the _____ transformation.

    A. Galilean

    B. Lorentz

    C. Euclidean

    D. Laplace

65. A baseball is thrown with an initial velocity of 30 m/s at an angle of 45°. Neglecting air resistance, how far away will the ball land?

    A. 92 m

    B. 78 m

    C. 65 m

    D. 46 m

66. If one sound is ten decibels louder than another, the ratio of the intensity of the first to the second is

    A. 20:1

    B. 10:1

    C. 1:1

    D. 1:10

67. A wave has speed 60 m/s and wavelength 30,000 m. What is the frequency of the wave?

    A. 2.0×10^{-3} Hz

    B. 60 Hz

    C. 5.0×10^{2} Hz

    D. 1.8×10^{6} Hz

68. An electromagnetic wave propagates through a vacuum. Independent of its wavelength, it will move with constant

   A. Acceleration

   B. Velocity

   C. Induction

   D. Sound

69. A wave generator is used to create a succession of waves. The rate of wave generation is one every 0.33 seconds. The period of these waves is

   A. 2.0 seconds

   B. 1.0 seconds

   C. 0.33 seconds

   D. 3.0 seconds

70. In a fission reactor, heavy water

   A. Cools off neutrons to control temperature

   B. Moderates fission reactions

   C. Initiates the reaction chain

   D. Dissolves control rods

71. Heat transfer by electromagnetic waves is termed

   A. Conduction

   B. Convection

   C. Radiation

   D. Phase Change

72. Solids expand when heated because

A. Molecular motion causes expansion

B. PV = nRT

C. Magnetic forces stretch the chemical bonds

D. All material is effectively fluid

73. Gravitational force at the earth's surface causes

A. All objects to fall with equal acceleration, ignoring air resistance

B. Some objects to fall with constant velocity, ignoring air resistance

C. A kilogram of feathers to float at a given distance above the earth

D. Aerodynamic objects to accelerate at an increasing rate

74. An office building entry ramp uses the principle of which simple machine?

A. Lever

B. Pulley

C. Wedge

D. Inclined Plane

75. The velocity of sound is greatest when traveling through

A. Water

B. Steel

C. Alcohol

D. Air

76. All of the following phenomena are considered "refractive effects" except for

    A. The red shift

    B. Total internal reflection

    C. Lens dependent image formation

    D. Snell's Law

77. Static electricity generation occurs by

    A. Telepathy

    B. Friction

    C. Removal of heat

    D. Evaporation

78. The wave phenomenon of polarization applies only to

    A. Longitudinal waves

    B. Transverse waves

    C. Sound

    D. Light

79. A force is given by the vector 5 N x + 3 N y (where x and y are the unit vectors for the x- and y- axes, respectively). This force is applied to move a 10 kg object 5 m, in the x direction. How much work was done?

    A. 250 J

    B. 400 J

    C. 40 J

    D. 25 J

80. A satellite is in a circular orbit above the earth. Which statement is false?

    A. An external force causes the satellite to maintain orbit.

    B. The satellite's inertia causes it to maintain orbit.

    C. The satellite is accelerating toward the earth.

    D. The satellite's velocity and acceleration are not in the same direction.

QUESTIONS AND ANSWERS
WITH EXPLANATIONS

1. A projectile with a mass of 1.0 kg has a muzzle velocity of 1500.0 m/s when it is fired from a cannon with a mass of 500.0 kg. If the cannon slides on a frictionless track, it will recoil with a velocity of _____ m/s.

   A. 2.4

   B. 3.0

   C. 3.5

   D. 1500

Answer:

B. 3.0

To solve this problem, apply Conservation of Momentum to the cannon-projectile system. The system is initially at rest, with total momentum of 0 kg m/s. Since the cannon slides on a frictionless track, we can assume that the net momentum stays the same for the system. Therefore, the momentum forward (of the projectile) must equal the momentum backward (of the cannon). Thus:

$p_{projectile} = p_{cannon}$

$m_{projectile}\ v_{projectile} = m_{cannon}\ v_{cannon}$

$(1.0\ kg)(1500.0\ m/s) = (500.0\ kg)(x)$

$x = 3.0\ m/s$

Only answer (B) matches these calculations.

2. The weight of an object on the earth's surface is designated x. When it is two earth's radii from the surface of the earth, its weight will be

   A. $x/4$

   B. $x/9$

   C. $4x$

   D. $16x$

Answer:

B. $x/9$

To solve this problem, apply the universal Law of Gravitation to the object and Earth:
$F_{gravity} = (GM_1M_2)/R^2$
Because the force of gravity varies with the square of the radius between the objects, the force (or weight) on the object will be decreased by the square of the multiplication factor on the radius. Note that the object on Earth's surface is *already* at one radius from Earth's center. Thus, when it is two radii from Earth's surface, it is three radii from Earth's center. R^2 is then nine, so the weight is $x/9$.
Only answer (B) matches these calculations.

3. **If a force of magnitude *F* gives a mass *M* an acceleration *A*, then a force *3F* would give a mass *3M* an acceleration**

   A. A

   B. $12A$

   C. $A/2$

   D. $6A$

Answer:

A. A

To solve this problem, apply Newton's Second Law, which is also implied by the first part of the problem:
Force = (Mass)(Acceleration)
$F = MA$
Then apply the same law to the second case, and isolate the unknown:
$3F = 3M x$
$x = (3F)/(3M)$
$x = F/M$
$x = A$ (by substituting from our first equation)
Only answer (A) matches these calculations.

4. A car (mass m_1) is driving at velocity v, when it smashes into an unmoving car (mass m_2), locking bumpers. Both cars move together at the same velocity. The common velocity will be given by

   A. m_1v/m_2

   B. m_2v/m_1

   C. $m_1v/(m_1 + m_2)$

   D. $(m_1 + m_2)v/m_1$

Answer:

C. $m_1v/(m_1 + m_2)$

In this problem, there is an inelastic collision, so the best method is to assume that momentum is conserved. (Recall that momentum is equal to the product of mass and velocity.)
Therefore, apply Conservation of Momentum to the two-car system:
Momentum at Start = Momentum at End
(Mom. of Car 1) + (Mom. of Car 2) = (Mom. of 2 Cars Coupled)
$m_1v + 0 = (m_1 + m_2)x$
$x = m_1v/(m_1 + m_2)$
Only answer (C) matches these calculations.

Watch out for the other answers, because errors in algebra could lead to a match with incorrect answer (D), and assumption of an elastic collision could lead to a match with incorrect answer (A).

5. When acceleration is plotted versus time, the area under the graph represents

   A. Time

   B. Distance

   C. Velocity

   D. Acceleration

Answer:

C. Velocity

The area under a graph will have units equal to the product of the units of the two axes. (To visualize this, picture a graphed rectangle with its area equal to length times width.)
Therefore, multiply units of acceleration by units of time:
$(\text{length/time}^2)(\text{time})$
This equals length/time, i.e. units of velocity.

6. **An inclined plane is tilted by gradually increasing the angle of elevation θ, until the block will slide down at a constant velocity. The coefficient of friction, μ_k, is given by**

    A. cos θ

    B. sin θ

    C. cosecant θ

    D. tangent θ

Answer:

D. tangent θ

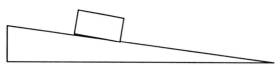

When the block moves, its force upstream (due to friction) must equal its force downstream (due to gravity).

The friction force is given by
$F_f = \mu_k\, N$
where μ_k is the friction coefficient and N is the normal force.

Using similar triangles, the gravity force is given by
$F_g = mg \sin θ$
and the normal force is given by
$N = mg \cos θ$

When the block moves at constant velocity, it must have zero net force, so set equal the force of gravity and the force due to friction:

$F_f = F_g$

$\mu_k \, mg \cos \theta = mg \sin \theta$

$\mu_k = \tan \theta$

Answer (D) is the only appropriate choice in this case.

7. **Use the information on heats below to solve this problem. An ice block at 0° Celsius is dropped into 100 g of liquid water at 18° Celsius. When thermal equilibrium is achieved, only liquid water at 0° Celsius is left. What was the mass, in grams, of the original block of ice?**

    Given: **Heat of fusion of ice = 80 cal/g**
                    Heat of vaporization of ice = 540 cal/g
                    Specific Heat of ice = 0.50 cal/g°C
                    Specific Heat of water = 1 cal/g°C

    A. 2.0

    B. 5.0

    C. 10.0

    D. 22.5

Answer:

D. 22.5

    To solve this problem, apply Conservation of Energy to the ice-water system. Any gain of heat to the melting ice must be balanced by loss of heat in the liquid water. Use the two equations relating temperature, mass, and energy:
Q = m C ΔT (for heat loss/gain from change in temperature)
Q = m L (for heat loss/gain from phase change)
where Q is heat change; m is mass; C is specific heat; ΔT is change in temperature; L is heat of phase change (in this case, melting, also known as "fusion").

Then

$Q_{ice\ to\ water} = Q_{water\ to\ ice}$

(Note that the ice only melts; it stays at 0° Celsius—otherwise, we would have to include a term for warming the ice as well. Also the information on the heat of vaporization for water is irrelevant to this problem.)

m L = m C ΔT

x (80 cal/g) = 100g 1cal/g°C 18°C

x (80 cal/g) = 1800 cal

x = 22.5 g

Only answer (D) matches this result.

8. **The combination of overtones produced by a musical instrument is known as its**

A. Timbre

B. Chromaticity

C. Resonant Frequency

D. Flatness

Answer:

A. Timbre

To answer this question, you must know some basic physics vocabulary. "Timbre" is the combination of tones that make a sound unique, beyond its pitch and volume. (For instance, consider the same note played at the same volume, but by different instruments.) Answer (A) is therefore the only appropriate choice. "Resonant Frequency" is relevant to music, because the resonant frequency of a wave will give the dominant sound tone. "Chromaticity" is an analogous word to "timbre," but it describes color tones. "Flatness" is unrelated, and incorrect.

9. **A long copper bar has a temperature of 60°C at one end and 0°C at the other. The bar reaches thermal equilibrium (barring outside influences) by the process of heat**

   A. Fusion

   B. Convection

   C. Conduction

   D. Microwaving

Answer:

C. Conduction

To answer this question, recall the different methods of heat transfer. (Note that since the bar is warm at one end and cold at the other, heat must transfer through the bar from warm to cold, until temperature is equalized.) "Convection" is the heat transfer via fluid currents. "Conduction" is the heat transfer via connected solid material. "Fusion" and "Microwaving" are not methods of heat transfer. Therefore the only appropriate answer is (C).

10. **The First Law of Thermodynamics takes the form dU = dW when the conditions are**

    A. Isobaric

    B. Isochloremic

    C. Isothermal

    D. Adiabatic

Answer:

D. Adiabatic

To answer this question, recall the First Law of Thermodynamics:
Change in Internal Energy = Work Done + Heat Added
dU = dW + dQ

Thus in the form we are given, dQ has been set to zero, i.e. there is no heat added. "Adiabatic" refers to a case where there is no heat exchange with surroundings, so answer (D) is the appropriate choice. "Isobaric" means at a constant pressure, "Isothermal" means at a constant temperature, and "Isochloremic" is an imaginary word, as far as I can tell.

It might be tempting to choose "Isothermal," thinking that no heat added would require the same temperature. However, work and internal energy changes can change temperature within the system analyzed, even when no heat is exchanged with the surroundings.

11. **Given a vase full of water, with holes punched at various heights. The water squirts out of the holes, achieving different distances before hitting the ground. Which of the following accurately describes the situation?**

   A. Water from higher holes goes farther, due to Pascal's Principle.

   B. Water from higher holes goes farther, due to Bernoulli's Principle.

   C. Water from lower holes goes farther, due to Pascal's Principle.

   D. Water from lower holes goes farther, due to Bernoulli's Principle.

Answer:

D. Water from lower holes goes farther, due to Bernoulli's Principle.

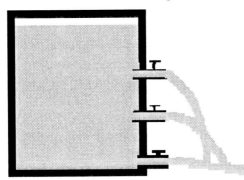

To answer this question, consider the pressure on the water in the vase. The deeper the water, the higher the pressure. Thus, when a hole is punched, the water stream will achieve higher velocity as it equalizes to atmospheric pressure. The lower streams will therefore travel farther before hitting the ground. This eliminates answers (A) and (B). Then recall that Pascal's Principle provides for immediate pressure changes throughout a fluid, while Bernoulli's Principle translates pressure, velocity, and height energy into each other. In this case, the pressure energy is being transformed into velocity energy, and Bernoulli's Principle applies. Therefore, the only appropriate answer is (D).

12. A stationary sound source produces a wave of frequency F. An observer at position A is moving toward the horn, while an observer at position B is moving away from the horn. Which of the following is true?

   A. $F_A < F < F_B$

   B. $F_B < F < F_A$

   C. $F < F_A < F_B$

   D. $F_B < F_A < F$

Answer:

B. $F_B < F < F_A$

To answer this question, recall the Doppler Effect. As a moving observer approaches a sound source, s/he intercepts wave fronts sooner than if s/he were standing still. Therefore, the wave fronts seem to be coming more frequently. Similarly, as an observer moves away from a sound source, the wave fronts take longer to reach him/her. Therefore, the wave fronts seem to be coming less frequently. Because of this effect, the frequency at B will seem lower than the original frequency, and the frequency at A will seem higher than the original frequency. The only answer consistent with this is (B). Note also, that even if you weren't sure of which frequency should be greater/smaller, you could still reason that A and B should have opposite effects, and be able to eliminate answer choices (C) and (D).

13. **The electric force in Newtons, on two small objects (each charged to −10 microCoulombs and separated by 2 meters) is**

   A. 1.0

   B. 9.81

   C. 31.0

   D. 0.225

Answer:

D. 0.225

To answer this question, use Coulomb's Law, which gives the electric force between two charged particles:
$F = k\ Q_1 Q_2 / r^2$

Then our unknown is F, and our knowns are:
$k = 9.0 \times 10^9\ Nm^2/C^2$
$Q_1 = Q_2 = -10 \times 10^{-6}\ C$
$r = 2\ m$

Therefore
$F = (9.0 \times 10^9)(-10 \times 10^{-6})(-10 \times 10^{-6})/(2^2)$ N
$F = 0.225\ N$

This is compatible only with answer (D).

14. A 10 ohm resistor and a 50 ohm resistor are connected in parallel. If the current in the 10 ohm resistor is 5 amperes, the current (in amperes) running through the 50 ohm resistor is

    A. 1

    B. 50

    C. 25

    D. 60

Answer:

    A. 1

To answer this question, use Ohm's Law, which relates voltage to current and resistance:
$V = IR$
where V is voltage; I is current; R is resistance.

We also use the fact that in a parallel circuit, the voltage is the same across the branches.

Because we are given that in one branch, the current is 5 amperes and the resistance is 10 ohms, we deduce that the voltage in this circuit is their product, 50 volts (from $V = IR$).

We then use $V = IR$ again, this time to find I in the second branch. Because V is 50 volts, and R is 50 ohm, we calculate that I has to be 1 ampere.

This is consistent only with answer (A).

15. Fahrenheit and Celsius thermometers have the same temperature reading at

    A. 100 degrees

    B. - 40 degrees

    C. Absolute Zero

    D. 40 degrees

Answer:

B. - 40 degrees

To answer this question, use the relationship between Fahrenheit and Celsius temperature scales:
F = 9/5 C + 32

Then, in a case where both °F and °C are equal, F = C, so
C = 9/5 C + 32
- 4/5 C = 32
C = - 40

Only answer (B) is consistent with this result.

16. **Which of the following apparatus can be used to measure the wavelength of a sound produced by a tuning fork?**

A. A glass cylinder, some water, and iron filings

B. A glass cylinder, a meter stick, and some water

C. A metronome and some ice water

D. A comb and some tissue

Answer:

B. A glass cylinder, a meter stick, and some water

To answer this question, recall that a sound will be amplified if it is reflected back to cause positive interference. This is the principle behind musical instruments that use vibrating columns of air to amplify sound (e.g. a pipe organ). Therefore, presumably a person could put varying amounts of water in the cylinder, and hold the vibrating tuning fork above the cylinder in each case. If the tuning fork sound is amplified when put at the top of the column, then the length of the air space would be an integral multiple of the sound's wavelength. This experiment is consistent with answer (B). Although the experiment would be tedious, none of the other options for materials suggest a better alternative.

17. When the current flowing through a fixed resistance is doubled, the amount of heat generated is

    A. Quadrupled

    B. Doubled

    C. Multiplied by pi

    D. Halved

Answer:

    A. Quadrupled

To answer this question, recall that heat generated will occur because of the power of the circuit (power is energy per time). For a circuit with a fixed resistance:
$P = I\,V$
where P is power; I is current; V is voltage. Then use Ohm's Law:
$V = I\,R$
where V is voltage; I is current; R is resistance, and substitute:
$P = I^2\,R$
and so the doubling of the current I will lead to a quadrupling of the power, and therefore the a quadrupling of the heat.

This is consistent only with answer (A). If you weren't sure of the equations, you could still deduce that with more current, there would be more heat generated, and therefore eliminate answer choice (D) in any case.

18. The current induced in a coil is defined by which of the following laws?

    A. Lenz's Law

    B. Burke's Law

    C. The Law of Spontaneous Combustion

    D. Snell's Law

Answer:

A. Lenz's Law

Lenz's Law states that an induced electromagnetic force always gives rise to a current whose magnetic field opposes the original flux change. There is no relevant "Snell's Law," "Burke's Law," or "Law of Spontaneous Combustion" in electromagnetism. (In fact, only Snell's Law is a real law of these three, and it refers to refracted light.) Therefore, the only appropriate answer is (A).

19. If an object is 15 cm from a convex lens whose focal length is 10 cm, the image is:

A. Virtual and upright

B. Real and inverted

C. Larger than the object

D. Smaller than the object

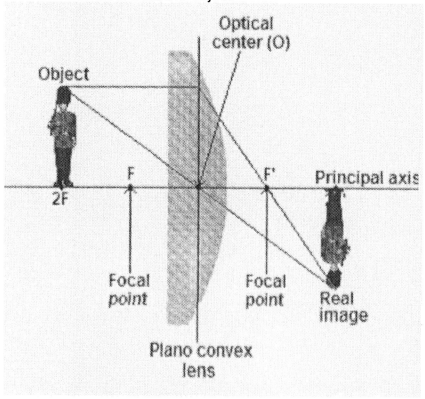

Answer:

B. Real and inverted

To solve this problem, draw a lens diagram with the lens, focal length, and image size.

The ray from the top of the object straight to the lens is focused through the far focus point; the ray from the top of the object through the near focus goes straight through the lens; the ray from the top of the object through the center of the lens continues. These three meet to form the "top" of the image, which is therefore real and inverted. This is consistent only with answer (B).

20. A cooking thermometer in an oven works because the metals it is composed of have different

A. Melting points

B. Heat convection

C. Magnetic fields

D. Coefficients of expansion

Answer:

D. Coefficients of expansion

A thermometer of the type that can withstand oven temperatures works by having more than one metal strip. These strips expand at different rates with temperature increases, causing the dial to register the new temperature. This is consistent only with answer (D). If you did not know how an oven thermometer works, you could still omit the incorrect answers: It is unlikely that the metals in a thermometer would melt in the oven to display the temperature; the magnetic fields would not be useful information in this context; heat convection applies in fluids, not solids.

21. In an experiment where a brass cylinder is transferred from boiling water into a beaker of cold water with a thermometer in it, we are measuring

    A. Fluid viscosity

    B. Heat of fission

    C. Specific heat

    D. Nonspecific heat

Answer:

C. Specific heat

In this question, we consider an experiment to measure temperature change of water (with the thermometer) as the cylinder cools and the water warms. This information can be used to calculate heat changes, and therefore specific heat. Therefore, (C) is the correct answer. Even if you were unable to deduce that specific heat is being measured, you could eliminate the other answer choices: viscosity cannot be measured with a thermometer; fission takes place at much higher temperatures than this experiment, and under quite different conditions; there is no such thing as "nonspecific heat".

22. A temperature change of 40 degrees Celsius is equal to a change in Fahrenheit degrees of

    A. 40

    B. 20

    C. 72

    D. 112

Answer:

C. 72

To answer this question, recall the equation for Celsius and Fahrenheit:
°F = 9/5 °C + 32

Therefore, whatever temperature difference occurs in °C, it is multiplied by a factor of 9/5 to get the new °F measurement:
new°F = 9/5(old°C + 40) + 32
(whereas old°F = 9/5(old°C) + 32)

Therefore the difference between the old and new temperatures in Fahrenheit is 9/5 of 40, or 72 degrees. This is consistent only with answer (C).

23. **The number of calories required to raise the temperature of 40 grams of water at 30°C to steam at 100°C is**

   A. 7500

   B. 23,000

   C. 24,400

   D. 30,500

Answer:

C. 24,400

To answer this question, apply the equations for heat transfer due to temperature and phase changes:
$Q = mC\Delta T + mL$
where Q is heat; m is mass; C is specific heat; ΔT is temperature change; L is heat of phase change.

In this problem, we are trying to find Q, and we are given:
m = 40 g
C = 1 cal/g°C for water (this should be memorized)
ΔT = 70 °C
L = 540 cal/g for liquid to gas change in water (this should be memorized)

thus Q = (40 g)(1 cal/g°C)(70 °C) + (40 g)(540 cal/g)
Q = 24,400 cal
This is consistent only with answer (C).

24. The boiling point of water on the Kelvin scale is closest to

    A. 112 K

    B. 212 K

    C. 373 K

    D. 473 K

Answer:

C. 373 K

To answer this question, recall that Kelvin temperatures are equal to Celsius temperatures plus 273.15. Since water boils at 100°C under standard conditions, it will boil at 373.15 K. This is consistent only with answer (C).

25. The kinetic energy of an object is _____ proportional to its _____.

    A. Inversely…inertia

    B. Inversely…velocity

    C. Directly…mass

    D. Directly…time

Answer:

C. Directly…mass

To answer this question, recall that kinetic energy is equal to one-half of the product of an object's mass and the square of its velocity:
$KE = \frac{1}{2} m v^2$

Therefore, kinetic energy is directly proportional to mass, and the answer is (C). Note that although kinetic energy is associated with both velocity and momentum (a measure of inertia), it is not *inversely* proportional to either one.

26. A hollow conducting sphere of radius R is charged with a total charge Q. What is the magnitude of the electric field at a distance r (given r<R) from the center of the sphere? (k is the electrostatic constant)

    A. 0

    B. $k \, Q/R^2$

    C. $k \, Q/(R^2 - r^2)$

    D. $k \, Q/(R - r)^2$

Answer:

A. 0

You may be tempted to use the equation for the electric field:
$E = F/Q$ (E = electric field; F = electric force; Q = charge)

and the Coulomb's Law expression for electric force:
$F = k \, Q_1 Q_2/r^2$ (k = constant; Q_1 and Q_2 = charges; r = distance apart),

which usually would give
$E = k \, Q/(R-r)^2$ in a similar context.

However, this question addresses a special case, i.e. a hollow conductor. Inside a hollow conductor, no electric field exists, because if there were an electric field inside, the conductor's free electrons would be forced to move (by the electric force) until the electric field became zero. Therefore, the only correct answer is (A).

27. A quantum of light energy is called a

    A. Dalton

    B. Photon

    C. Curie

    D. Heat Packet

Answer:

B. Photon

The smallest "packet" (quantum) of light energy is a photon. "Heat Packet" does not have any relevant meaning, and while "Dalton" and "Curie" have other meanings, they are not connected to light. Therefore, only (B) is a correct answer.

28. The following statements about sound waves are true *except*

A. Sound travels faster in liquids than in gases.

B. Sound waves travel through a vacuum.

C. Sound travels faster through solids than liquids.

D. Ultrasound can be reflected by the human body.

Answer:

B. Sound waves travel through a vacuum.

Sound waves require a medium to travel. The sound wave agitates the material, and this occurs fastest in solids, then liquids, then gases. Ultrasound waves are reflected by parts of the body, and this is useful in medical imaging. Therefore, the only correct answer is (B).

29. The greatest number of 100 watt lamps that can be connected in parallel with a 120 volt system without blowing a 5 amp fuse is

A. 24

B. 12

C. 6

D. 1

Answer:

C. 6

To solve fuse problems, you must add together all the drawn current in the parallel branches, and make sure that it is less than the fuse's amp measure. Because we know that electrical power is equal to the product of current and voltage, we can deduce that:
$I = P/V$ (I = current (amperes); P = power (watts); V = voltage (volts))

Therefore, for each lamp, the current is 100/120 amperes, or 5/6 ampere. The highest possible number of lamps is thus six, because six lamps at 5/6 ampere each adds to 5 amperes; more will blow the fuse.

This is consistent only with answer (C).

30. **A monochromatic ray of light passes from air to a thick slab of glass (n = 1.41) at an angle of 45° from the normal. At what angle does it leave the air/glass interface?**

   A. 45°

   B. 30°

   C. 15°

   D. 55°

Answer:

B. 30°

To solve this problem use Snell's Law:
$n_1 \sin\theta_1 = n_2 \sin\theta_2$ (where n_1 and n_2 are the indeces of refraction and θ_1 and θ_2 are the angles of incidence and refraction).

Then, since the index of refraction for air is 1.0, we deduce:
$1 \sin 45° = 1.41 \sin x$
$x = \sin^{-1} ((1/1.41) \sin 45°)$
$x = 30°$

This is consistent only with answer (B). Also, note that you could eliminate answers (A) and (D) in any case, because the refracted light will have to bend at a smaller angle when entering glass.

31. The magnitude of a force is

    A. Directly proportional to mass and inversely to acceleration

    B. Inversely proportional to mass and directly to acceleration

    C. Directly proportional to both mass and acceleration

    D. Inversely proportional to both mass and acceleration

Answer:

    C. Directly proportional to both mass and acceleration

To solve this problem, recall Newton's 2nd Law, i.e. net force is equal to mass times acceleration. Therefore, the only possible answer is (C).

32. A semi-conductor allows current to flow

    A. Never

    B. Always

    C. As long as it stays below a maximum temperature

    D. When a minimum voltage is applied

Answer:

    D. When a minimum voltage is applied

To answer this question, recall that semiconductors do not conduct as well as conductors (eliminating answer (B)), but they conduct better than insulators (eliminating answer (A)). Semiconductors can conduct better when the temperature is higher (eliminating answer (C)), and their electrons move most readily under a potential difference. Thus the answer can only be (D).

33. One reason to use salt for melting ice on roads in the winter is that

    A. Salt lowers the freezing point of water.

    B. Salt causes a foaming action, which increases traction.

    C. Salt is more readily available than sugar.

    D. Salt increases the conductivity of water.

Answer:

    A. Salt lowers the freezing point of water.

In answering this question, you may recall that salt is used for road traction because it has large particles to increase traction (analogous to sand). This is true, but salt also has the potential to lower the freezing point of water, thus melting the ice with which it has contact. This is consistent with answer (A). Answer (B) is untrue, and the other two choices, while usually true, are irrelevant in this case.

34. Automobile mirrors that have a sign, "objects are closer than they appear" say so because

    A. The real image of an obstacle, through a converging lens, appears farther away than the object.

    B. The real or virtual image of an obstacle, through a converging mirror, appears farther away than the object.

    C. The real image of an obstacle, through a diverging lens, appears farther away than the object.

    D. The virtual image of an obstacle, through a diverging mirror, appears farther away than the object.

Answer:

D. The virtual image of an obstacle, through a diverging mirror, appears farther away than the object.

To answer this question, first eliminate answer choices (A) and (C), because we have a mirror, not a lens. Then draw ray diagrams for diverging (convex) and converging (concave) mirrors, and note that because the focal point of a diverging mirror is behind the surface, the image is smaller than the object. This creates the illusion that the object is farther away, and therefore (D) is the correct answer.

35. **A gas maintained at a constant pressure has a specific heat which is greater than its specific heat when maintained at constant volume, because**

    A. The Coefficient of Expansion changes.

    B. The gas enlarges as a whole.

    C. Brownian motion causes random activity.

    D. Work is done to expand the gas.

Answer:

D. Work is done to expand the gas.

To answer this question, recall that the specific heat is a measure of how much energy it takes to raise the temperature of a given mass of gas. Thus, you can reason that when a gas is maintained at constant pressure, some energy is used to expand the volume of the gas, and less is left for temperature changes. In fact, this is the case, and (D) is the correct answer. If you were not able to figure that out, you could still eliminate the other answers, because they are not strictly relevant to a change in specific heat.

36. Consider the shear modulus of water, and that of mercury. Which of the following is true?

    A. Mercury's shear modulus indicates that it is the only choice of fluid for a thermometer.

    B. The shear modulus of each of these is zero.

    C. The shear modulus of mercury is higher than that of water.

    D. The shear modulus of water is higher than that of mercury.

Answer:

    B. The shear modulus of each of these is zero.

    To answer this question, recall that shear modulus is meaningful only for solids, and that liquids, instead, have a bulk modulus. The only reasonable answer is therefore (B).

37. A skateboarder accelerates down a ramp, with constant acceleration of two meters per second squared, from rest. The distance in meters, covered after four seconds, is

    A. 10

    B. 16

    C. 23

    D. 37

Answer:

    B. 16

    To answer this question, recall the equation relating constant acceleration to distance and time:
$x = \frac{1}{2} a t^2 + v_0 t + x_0$ where x is position; a is acceleration; t is time; v_0 and x_0 are initial velocity and position (both zero in this case)

    thus, to solve for x:
$x = \frac{1}{2} (2 \text{ m/s}^2) (4^2 \text{s}^2) + 0 + 0$
$x = 16 \text{ m}$

    This is consistent only with answer (B).

38. Which of the following units is not used to measure torque?

    A. slug ft

    B. lb ft

    C. N m

    D. dyne cm

Answer:

    A. slug ft

To answer this question, recall that torque is always calculated by multiplying units of force by units of distance. Therefore, answer (A), which is the product of units of mass and units of distance, must be the choice of incorrect units. Indeed, the other three answers all could measure torque, since they are of the correct form. It is a good idea to review "English Units" before the teacher test, because they are occasionally used in problems.

39. In a nuclear pile, the control rods are composed of

    A. Boron

    B. Einsteinium

    C. Isoptocarpine

    D. Phlogiston

Answer:

    A. Boron

Nuclear plants use control rods made of boron or cadmium, to absorb neutrons and maintain "critical" conditions in the reactor. However, if you did not know that, you could still eliminate choice (D), because "phlogiston" is the word for the imaginary element in an ancient structure of earth-air-water-fire.

40. All of the following use semi-conductor technology, except a(n):

    A. Transistor

    B. Diode

    C. Capacitor

    D. Operational Amplifier

Answer:

    C. Capacitor

Semi-conductor technology is used in transistors and operational amplifiers, and diodes are the basic unit of semi-conductors. Therefore the only possible answer is (C), and indeed a capacitor does not require semi-conductor technology.

41. Ten grams of a sample of a radioactive material (half-life = 12 days) were stored for 48 days and re-weighed. The new mass of material was

    A. 1.25 g

    B. 2.5 g

    C. 0.83 g

    D. 0.625 g

Answer:

    D. 0.625 g

To answer this question, note that 48 days is four half-lives for the material. Thus, the sample will degrade by half four times. At first, there are ten grams, then (after the first half-life) 5 g, then 2.5 g, then 1.25 g, and after the fourth half-life, there remains 0.625 g. You could also do the problem mathematically, by multiplying ten times $(\frac{1}{2})^4$, i.e. $\frac{1}{2}$ for each half-life elapsed.

42. When a radioactive material emits an alpha particle only, its atomic number will

    A. Decrease

    B. Increase

    C. Remain unchanged

    D. Change randomly

Answer:

A. Decrease

To answer this question, recall that in alpha decay, a nucleus emits the equivalent of a Helium atom. This includes two protons, so the original material changes its atomic number by a decrease of two.

43. The sun's energy is produced primarily by

    A. Fission

    B. Explosion

    C. Combustion

    D. Fusion

Answer:

D. Fusion

To answer this question, recall that in stars (such as the sun), fusion is the main energy-producing occurrence. Fission, explosion, and combustion all release energy in other contexts, but they are not the right answers here.

44. A crew is on-board a spaceship, traveling at 60% of the speed of light with respect to the earth. The crew measures the length of their ship to be 240 meters. When a ground-based crew measures the apparent length of the ship, it equals

    A. 400 m

    B. 300 m

    C. 240 m

    D. 192 m

Answer:

D. 192 m

To answer this question, recall that a moving object's size seems contracted to the stationary observer, according to the equation:
Length Observed = (Actual Length) $(1 - v^2/c^2)^{1/2}$
where v is speed of motion, and c is speed of light.

Therefore, in this case,
Length Observed = (240 m) $(1 - 0.6^2)^{1/2}$
Length Observed = (240 m) (0.8) = 192 m

This is consistent only with answer (D). If you were unsure of the equation, you could still reason that because of length contraction (the flip side of time dilation), you must choose an answer with a smaller length, and only (D) fits that description. Note that only the dimension in the direction of travel is contracted. (The length in this case.)

45. Which of the following pairs of elements are not found to fuse in the centers of stars?

    A. Oxygen and Helium

    B. Carbon and Hydrogen

    C. Beryllium and Helium

    D. Cobalt and Hydrogen

Answer:

D. Cobalt and Hydrogen

To answer this question, recall that fusion is possible only when the final product has more binding energy than the reactants. Because binding energy peaks near a mass number of around 56, corresponding to Iron, any heavier elements would be unlikely to fuse in a typical star. (In very massive stars, there may be enough energy to fuse heavier elements.) Of all the listed elements, only Cobalt is heavier than iron, so answer (D) is correct.

46. A calorie is the amount of heat energy that will

A. Raise the temperature of one gram of water from 14.5° C to 15.5° C.

B. Lower the temperature of one gram of water from 16.5° C to 15.5° C

C. Raise the temperature of one gram of water from 32° F to 33° F

D. Cause water to boil at two atmospheres of pressure.

Answer:

A. Raise the temperature of one gram of water from 14.5° C to 15.5° C.

The definition of a calorie is, "the amount of energy to raise one gram of water by one degree Celsius," and so answer (A) is correct. Do not get confused by the fact that 14.5° C seems like a random number. Also, note that answer (C) tries to confuse you with degrees Fahrenheit, which are irrelevant to this problem.

47. Bohr's theory of the atom was the first to quantize

A. Work

B. Angular Momentum

C. Torque

D. Duality

Answer:

B. Angular Momentum

Bohr was the first to quantize the angular momentum of electrons, as he combined Rutherford's planet-style model with his knowledge of emerging quantum theory. Recall that he derived a "quantum condition" for the single electron, requiring electrons to exist at specific energy levels.

48. A uniform pole weighing 100 grams, that is one meter in length, is supported by a pivot at 40 centimeters from the left end. In order to maintain static position, a 200 gram mass must be placed _____ centimeters from the left end.

    A. 10

    B. 45

    C. 35

    D. 50

Answer:

D. 50

In answering this question, do not be tricked into calculating the position of the mass to create balance on the pole's pivot. (This calculation, with equal torques, would lead to incorrect answer (B).) A careful read of the question reveals that we want to "maintain static position" i.e. keep the pole from moving. Because it is already tilted toward the right side, that side (or anywhere to the right of the 45 cm pivot balance answer) is the correct place to put the additional weight without causing the pole to move. Thus, only answer (D) can be correct.

49. A classroom demonstration shows a needle floating in a tray of water. This demonstrates the property of

    A. Specific Heat

    B. Surface Tension

    C. Oil-Water Interference

    D. Archimedes' Principle

Answer:

B. Surface Tension

To answer this question, note that the only information given is that the needle (a small object) floats on the water. This occurs because although the needle is denser than the water, the surface tension of the water causes sufficient resistance to support the small needle. Thus the answer can only be (B). Answer (A) is unrelated to objects floating, and while answers (C) and (D) could be related to water experiments, they are not correct in this case. There is no oil in the experiment, and Archimedes' Principle allows the equivalence of displaced volumes, which is not relevant here.

50. Two neutral isotopes of a chemical element have the same numbers of

    A. Electrons and Neutrons

    B. Electrons and Protons

    C. Protons and Neutrons

    D. Electrons, Neutrons, and Protons

Answer:

B. Electrons and Protons

To answer this question, recall that isotopes vary in their number of neutrons. (This fact alone eliminates answers (A), (C), and (D).) If you did not recall that fact, note that we are given that the two samples are of the same element, constraining the number of protons to be the same in each case. Then, use the fact that the samples are neutral, so the number of electrons must exactly balance the number of protons in each case. The only correct answer is thus (B).

51. A mass is moving at constant speed in a circular path. Choose the true statement below:

    A. Two forces in equilibrium are acting on the mass.

    B. No forces are acting on the mass.

    C. One centripetal force is acting on the mass.

    D. One force tangent to the circle is acting on the mass.

Answer:

C. One centripetal force is acting on the mass.

To answer this question, recall that by Newton's 2nd Law, F = ma. In other words, force is mass times acceleration. Furthermore, acceleration is any change in the velocity vector—whether in size or direction. In circular motion, the direction of velocity is constantly changing. Therefore, there must be an unbalanced force on the mass to cause that acceleration. This eliminates answers (A) and (B) as possibilities. Recall then that the mass would ordinarily continue traveling tangent to the circle (by Newton's 1st Law). Therefore, the force must be to cause the turn, i.e. a centripetal force. Thus, the answer can only be (C).

52. A light bulb is connected in series with a rotating coil within a magnetic field. The brightness of the light may be increased by any of the following except:

    A. Rotating the coil more rapidly.

    B. Using more loops in the coil.

    C. Using a different color wire for the coil.

    D. Using a stronger magnetic field.

Answer:

C. Using a different color wire for the coil.

To answer this question, recall that the rotating coil in a magnetic field generates electric current, by Faraday's Law. Faraday's Law states that the amount of emf generated is proportional to the rate of change of magnetic flux through the loop. This increases if the coil is rotated more rapidly (A), if there are more loops (B), or if the magnetic field is stronger (D). Thus, the only answer to this question is (C).

53. **The use of two circuits next to each other, with a change in current in the primary circuit, demonstrates**

    A. Mutual current induction

    B. Dielectric constancy

    C. Harmonic resonance

    D. Resistance variation

Answer:

A. Mutual current induction

To answer this question, recall that changing current induces a change in magnetic flux, which in turn causes a change in current to oppose that change (Lenz's and Faraday's Laws). Thus, (A) is correct. If you did not remember that, note that harmonic resonance is irrelevant here (eliminating (C)), and there is no change in resistance in the circuits (eliminating (D)).

54. **A brick and hammer fall from a ledge at the same time. They would be expected to**

    A. Reach the ground at the same time

    B. Accelerate at different rates due to difference in weight

    C. Accelerate at different rates due to difference in potential energy

    D. Accelerate at different rates due to difference in kinetic energy

Answer:

A. Reach the ground at the same time

This is a classic question about falling in a gravitational field. All objects are acted upon equally by gravity, so they should reach the ground at the same time. (In real life, air resistance can make a difference, but not at small heights for similarly shaped objects.) In any case, weight, potential energy, and kinetic energy do not affect gravitational acceleration. Thus, the only possible answer is (A).

55. **The potential difference across a five Ohm resistor is five Volts. The power used by the resistor, in Watts, is**

    A. 1

    B. 5

    C. 10

    D. 20

Answer:

B. 5

To answer this question, recall the two relevant equations for potential difference and electric power:
$V = IR$ (where V is voltage; I is current; R is resistance)
$P = IV = I^2R$ (where P is power; I is current; R is resistance)

Thus, first calculate the current from the first equation:
$I = V/R = 1$ Ampere

And then use the second equation:
$P = I^2R = 5$ Watts

This is consistent only with answer (B).

56. An object two meters tall is speeding toward a plane mirror at 10 m/s. What happens to the image as it nears the surface of the mirror?

    A. It becomes inverted.

    B. The Doppler Effect must be considered.

    C. It remains two meters tall.

    D. It changes from a real image to a virtual image.

Answer:

C. It remains two meters tall.

Note that the mirror is a plane mirror, so the image is always a virtual image of the same size as the object. If the mirror were concave, then the image would be inverted until the object came within the focal distance of the mirror. The Doppler Effect is not relevant here. Thus, the only possible answer is (C).

57. The highest energy is associated with

    A. UV radiation

    B. Yellow light

    C. Infrared radiation

    D. Gamma radiation

Answer:

D. Gamma radiation

To answer this question, recall the electromagnetic spectrum. The highest energy (and therefore frequency) rays are those with the lowest wavelength, i.e. gamma rays. (In order of frequency from lowest to highest are: radio, microwave, infrared, red through violet visible light, ultraviolet, X-rays, gamma rays.) Thus, the only possible answer is (D). Note that even if you did not remember the spectrum, you could deduce that gamma radiation is considered dangerous and thus might have the highest energy.

58. The constant of proportionality between the energy and the frequency of electromagnetic radiation is known as the

    A. Rydberg constant

    B. Energy constant

    C. Planck constant

    D. Einstein constant

Answer:

C. Planck constant

Planck estimated his constant to determine the ratio between energy and frequency of radiation. The Rydberg constant is used to find the wavelengths of the visible lines on the hydrogen spectrum. The other options are not relevant options, and may not actually have physical meaning. Therefore, the only possible answer is (C).

59. A simple pendulum with a period of one second has its mass doubled. If the length of the string is quadrupled, the new period will be

    A. 1 second

    B. 2 seconds

    C. 3 seconds

    D. 5 seconds

Answer:

B. 2 seconds

To answer this question, recall that the period of a pendulum is given by:

$T = 2\pi (L/g)^{1/2}$ where T is period; L is length; g is gravitational acceleration (This is derived from balancing the forces and making trigonometric assumptions for small angles.)
Note that this equation is independent of mass, so that change is irrelevant. Since the length is quadrupled, and all other quantities on the right side of the equation are constant, the new period will be increased by a factor of two (the square root of four).
This is consistent only with answer (B).

60. A vibrating string's frequency is _____ proportional to the _____.

 A. Directly; Square root of the tension

 B. Inversely; Length of the string

 C. Inversely; Squared length of the string

 D. Inversely; Force of the plectrum

Answer:

A. Directly; Square root of the tension

To answer this question, recall that
$f = (n\,v) / (2\,L)$ where f is frequency; v is velocity; L is length

and

$v = (F_{tension} / (m / L))^{\frac{1}{2}}$ where $F_{tension}$ is tension; m is mass; others as above

so

$f = (n / 2\,L)\,((F_{tension} / (m / L))^{\frac{1}{2}}\,)$

indicating that frequency is directly proportional to the square root of the tension force. This is consistent only with answer (A). Note that in the final frequency equation, there is an inverse relationship with the square root of the length (after canceling like terms). This is not one of the options, however.

61. When an electron is "orbiting" the nucleus in an atom, it is said to posses an intrinsic spin (spin angular momentum). How many values can this spin have in any given electron?

 A. 1

 B. 2

 C. 3

 D. 8

Answer:

B. 2

> To answer this question, recall that electrons fill orbitals in pairs, and the two electrons in any pair have opposite spin from one another. Thus, (B) is correct. Note that answer (D) is trying to mislead you into thinking of the number of valence electrons in an atom.

62. Electrons are

A. More massive than neutrons

B. Positively charged

C. Neutrally charged

D. Negatively charged

Answer:

D. Negatively charged

> Electrons are negatively charged particles that have a tiny mass compared to protons and neutrons. Thus, answer (D) is the only correct alternative.

63. Rainbows are created by

A. Reflection, dispersion, and recombination

B. Reflection, resistance, and expansion

C. Reflection, compression, and specific heat

D. Reflection, refraction, and dispersion

Answer:

D. Reflection, refraction, and dispersion

> To answer this question, recall that rainbows are formed by light that goes through water droplets and is dispersed into its colors. This is consistent with both answers (A) and (D). Then note that refraction is important in bending the differently colored light waves, while recombination is not a relevant concept here. Therefore, the answer is (D).

64. In order to switch between two different reference frames in special relativity, we use the _____ transformation.

    A. Galilean

    B. Lorentz

    C. Euclidean

    D. Laplace

Answer:

B. Lorentz

The Lorentz transformation is the set of equations to scale length and time between inertial reference frames in special relativity, when velocities are close to the speed of light. The Galilean transformation is a parallel set of equations, used for 'classical' situations when velocities are much slower than the speed of light. Euclidean geometry is useful in physics, but not relevant here. Laplace transforms are a method of solving differential equations by using exponential functions. The correct answer is therefore (B).

65. A baseball is thrown with an initial velocity of 30 m/s at an angle of 45°. Neglecting air resistance, how far away will the ball land?

    A. 92 m

    B. 78 m

    C. 65 m

    D. 46 m

Answer:

A. 92 m

To answer this question, recall the equations for projectile motion:
$$y = \tfrac{1}{2} a\, t^2 + v_{0y}\, t + y_0$$
$$x = v_{0x}\, t + x_0$$
where x and y are horizontal and vertical position, respectively; t is time; a is acceleration due to gravity; v_{0x} and v_{0y} are initial horizontal and vertical velocity, respectively; x_0 and y_0 are initial horizontal and vertical position, respectively.

For our case:
x_0 and y_0 can be set to zero
both v_{0x} and v_{0y} are (using trigonometry) = $(\sqrt{2} / 2)$ 30 m/s
a = -9.81 m/s^2

We then use the vertical motion equation to find the time aloft (setting y equal to zero to find the solution for t):
$0 = \frac{1}{2} (-9.81 \text{ m/s}^2) t^2 + (\sqrt{2} / 2)$ 30 m/s t
Then solving, we find:
t = 0 s (initial set-up) or t = 4.324 s (time to go up and down)

Using t = 4.324 s in the horizontal motion equation, we find:
x = $((\sqrt{2} / 2)$ 30 m/s) (4.324 s)
x = 91.71 m

This is consistent only with answer (A).

66. **If one sound is ten decibels louder than another, the ratio of the intensity of the first to the second is**

    A. 20:1

    B. 10:1

    C. 1:1

    D. 1:10

Answer:

B. 10:1

To answer this question, recall that a decibel is defined as ten times the log of the ratio of sound intensities:
(decibel measure) = $10 \log (I / I_0)$ where I_0 is a reference intensity.

Therefore, in our case,
(decibels of first sound) = (decibels of second sound) + 10
$10 \log (I_1 / I_0) = 10 \log (I_2 / I_0) + 10$
$10 \log I_1 - 10 \log I_0 = 10 \log I_2 - 10 \log I_0 + 10$
$10 \log I_1 - 10 \log I_2 = 10$
$\log (I_1 / I_2) = 1$
$I_1 / I_2 = 10$

This is consistent only with answer (B).
(Be careful not to get the two intensities confused with each other.)

67. A wave has speed 60 m/s and wavelength 30,000 m. What is the frequency of the wave?

  A. 2.0×10^{-3} Hz

  B. 60 Hz

  C. 5.0×10^2 Hz

  D. 1.8×10^6 Hz

Answer:

A. 2.0×10^{-3} Hz

  To answer this question, recall that wave speed is equal to the product of wavelength and frequency. Thus:
  60 m/s = (30,000 m) (frequency)
  frequency = 2.0×10^{-3} Hz

  This is consistent only with answer (A).

68. An electromagnetic wave propagates through a vacuum. Independent of its wavelength, it will move with constant

  A. Acceleration

  B. Velocity

  C. Induction

  D. Sound

Answer:

B. Velocity

  Electromagnetic waves are considered always to travel at the speed of light, so answer (B) is correct. Answers (C) and (D) can be eliminated in any case, because induction is not relevant here, and sound does not travel in a vacuum.

69. A wave generator is used to create a succession of waves. The rate of wave generation is one every 0.33 seconds. The period of these waves is

    A. 2.0 seconds

    B. 1.0 seconds

    C. 0.33 seconds

    D. 3.0 seconds

Answer:

C. 0.33 seconds

The definition of a period is the length of time between wave crests. Therefore, when waves are generated one per 0.33 seconds, that same time (0.33 seconds) is the period. This is consistent only with answer (C). Do not be trapped into calculating the number of waves per second, which might lead you to choose answer (D).

70. In a fission reactor, heavy water

    A. Cools off neutrons to control temperature

    B. Moderates fission reactions

    C. Initiates the reaction chain

    D. Dissolves control rods

Answer:

B. Moderates fission reactions

In a nuclear reactor, heavy water is made up of oxygen atoms with hydrogen atoms called 'deuterium,' which contain two neutrons each. This allows the water to slow down (moderate) the neutrons, without absorbing many of them. This is consistent only with answer (B).

71. Heat transfer by electromagnetic waves is termed

    A. Conduction

    B. Convection

    C. Radiation

    D. Phase Change

Answer:

C. Radiation

To answer this question, recall the different ways that heat is transferred. Conduction is the transfer of heat through direct physical contact and molecules moving and hitting each other. Convection is the transfer of heat via density differences and flow of fluids. Radiation is the transfer of heat via electromagnetic waves (and can occur in a vacuum). Phase Change causes transfer of heat (though not of temperature) in order for the molecules to take their new phase. This is consistent, therefore, only with answer (C).

72. Solids expand when heated because

    A. Molecular motion causes expansion

    B. $PV = nRT$

    C. Magnetic forces stretch the chemical bonds

    D. All material is effectively fluid

Answer:

A. Molecular motion causes expansion

When any material is heated, the heat energy becomes energy of motion for the material's molecules. This increased motion causes the material to expand (or sometimes to change phase). Therefore, the answer is (A). Answer (B) is the ideal gas law, which gives a relationship between temperature, pressure, and volume for gases. Answer (C) is a red herring (misleading answer that is untrue). Answer (D) may or may not be true, but it is not the best answer to this question.

73. Gravitational force at the earth's surface causes

    A. All objects to fall with equal acceleration, ignoring air resistance

    B. Some objects to fall with constant velocity, ignoring air resistance

    C. A kilogram of feathers to float at a given distance above the earth

    D. Aerodynamic objects to accelerate at an increasing rate

Answer:

A. All objects to fall with equal acceleration, ignoring air resistance

Gravity acts to cause equal acceleration on all objects, though our atmosphere causes air resistance that slows some objects more than others. This is consistent only with answer (A). Answer (B) is incorrect, because ignoring air resistance leads to the result of constant acceleration, not zero acceleration. Answer (C) is incorrect because all objects (except tiny ones in which random Brownian motion is more significant than gravity) eventually fall due to gravity. Answer (D) is incorrect because it is not related to the constant acceleration due to gravity.

74. An office building entry ramp uses the principle of which simple machine?

    A. Lever

    B. Pulley

    C. Wedge

    D. Inclined Plane

Answer:

D. Inclined Plane

To answer this question, recall the definitions of the various simple machines. A ramp, which trades a longer traversed distance for a shallower slope, is an example of an Inclined Plane, consistent with answer (D). Levers and Pulleys act to change size and/or direction of an input force, which is not relevant here. Wedges apply the same force over a smaller area, increasing pressure—again, not relevant in this case.

75. The velocity of sound is greatest when traveling through

    A. Water

    B. Steel

    C. Alcohol

    D. Air

Answer:

B. Steel

Sound is a longitudinal wave, which means that it shakes its medium in a way that propagates as sound traveling. The speed of sound depends on both elastic modulus and density, but for a comparison of the above choices, the answer is always that sound travels faster through a solid like steel, than through liquids or gases. Thus, the answer is (B).

76. All of the following phenomena are considered "refractive effects" except for

    A. The red shift

    B. Total internal reflection

    C. Lens dependent image formation

    D. Snell's Law

Answer:

A. The red shift

Refractive effects are phenomena that are related to or caused by refraction. The red shift refers to the Doppler Effect as applied to light when galaxies travel away from observers. Total internal reflection is when light is totally reflected in a substance, with no refracted ray into the substance beyond (e.g. in fiber optic cables). It occurs because of the relative indices of refraction in the materials. Lens dependent image formation refers to making images depending on the properties (including index of refraction) of the lens. Snell's Law provides a mathematical relationship for angles of incidence and refraction. Therefore, the only possible answer is (A).

77. Static electricity generation occurs by

    A. Telepathy

    B. Friction

    C. Removal of heat

    D. Evaporation

Answer:

B. Friction

Static electricity occurs because of friction and electric charge build-up. There is no such thing as telepathy, and neither removal of heat nor evaporation are causes of static electricity. Therefore, the only possible answer is (B).

78. The wave phenomenon of polarization applies only to

    A. Longitudinal waves

    B. Transverse waves

    C. Sound

    D. Light

Answer:

B. Transverse waves

To answer this question, recall that polarization is when waves are screened so that they come out aligned in a certain direction. (To illustrate this, take two pairs of polarizing sunglasses, and note the light differences when rotating one lens over another. When the lenses are polarizing perpendicularly, no light gets through.) This applies only to transverse waves, which have wave parts to align. Light can be polarized, but it is not the only wave that can be. Thus, the correct answer is (B).

79. A force is given by the vector 5 N x + 3 N y (where x and y are the unit vectors for the x- and y- axes, respectively). This force is applied to move a 10 kg object 5 m, in the x direction. How much work was done?

    A. 250 J

    B. 400 J

    C. 40 J

    D. 25 J

Answer:

D. 25 J

To find out how much work was done, note that work counts only the force in the direction of motion. Therefore, the only part of the vector that we use is the 5 N in the x-direction. Note, too, that the mass of the object is not relevant in this problem. We use the work equation:
Work = (Force in direction of motion) (Distance moved)
Work = (5 N) (5 m)
Work = 25 J
This is consistent only with answer (D).

80. A satellite is in a circular orbit above the earth. Which statement is false?

    A. An external force causes the satellite to maintain orbit.

    B. The satellite's inertia causes it to maintain orbit.

    C. The satellite is accelerating toward the earth.

    D. The satellite's velocity and acceleration are not in the same direction.

Answer:

B. The satellite's inertia causes it to maintain orbit.

To answer this question, recall that in circular motion, an object's inertia tends to keep it moving straight (tangent to the orbit), so a centripetal force (leading to centripetal acceleration) must be applied. In this case, the centripetal force is gravity due to the earth, which keeps the object in motion. Thus, (A), (C), and (D) are true, and (B) is the only false statement.

Printed in the United States
52674LVS00005BA/23